A Puzzle a Day

A collection of mathematical problems
for every day of the school year

SOLVE INVESTIGATE & QUESTION
QUESTION SOLVE & INVESTIGATE
INVESTIGATE QUESTION & SOLVE
SOLVE INVESTIGATE & QUESTION
QUESTION SOLVE & INVESTIGATE
INVESTIGATE QUESTION & SOLVE
SOLVE INVESTIGATE & QUESTION
QUESTION SOLVE & INVESTIGATE
INVESTIGATE QUESTION & SOLVE
SOLVE INVESTIGATE & QUESTION
QUESTION SOLVE & INVESTIGATE
INVESTIGATE QUESTION & SOLVE

Vivien Lucas

I should like to thank all those who have helped me with the preparation of this book, in particular my friend and colleague Dot Folwell. She has worked through all the puzzles and has made many helpful suggestions. Also I should like to thank my pupils whose enthusiasm for my daily 'Mystery Number' competition gave me the inspiration to do this book.

Tarquin Publications
Suite 74,17Holywell Hill
AL1 1DT
www.tarquingroup.com

A Puzzle for Every Day of the School Year

There are roughly 180 days in the school year, hence 180 puzzles in this book. There are two similar questions on 90 different topics and my suggestion is that the second question on the topic can be offered about six months after the first. They are primarily intended for use in secondary schools, but with care many of the ideas could be used for younger children too.

The answers are short, indeed most are just a single number. The way these puzzles have been used in my school is for a new question to be displayed on the classroom door each morning and for the answers to be posted into a letter box during the day. Generally, solving the problem is a sufficient reward in itself but a small prize can be offered from time to time.

A further use for these puzzles is as 'fillers' to occupy a few spare minutes. The topics are not intended for direct use within the curriculum nor do they require much special knowledge. They are suitable for a wide range of interests and abilities.

I hope that you find this collection useful in helping to increase interest in mathematical thinking and vocabulary.

Vivien Lucas

Numerically the Same?

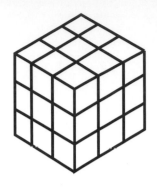

1a Find the number that is both
the surface area in square centimetres
and the volume in cubic centimetres
of the same cube.

1b Find the number which represents the
volume of a cube in cubic centimetres that is
numerically double the surface area in
square centimetres of the same cube.

What Multiple of 53 is This?

2a

It has three digits and is odd.
It reads the same when rotated through 180°.
It is a multiple of 53.

2b

It has three digits and is odd.
It is a multiple of 53.
When rotated through 180° it is an even
multiple of 53.

Which Perfect Square is This?

3a
It has three digits.
It is a perfect square.
If you swap the last two digits
it is still a perfect square.
The square root is prime.

3b
It has three digits.
It is a perfect square.
It is palindromic.
The sum of its digits is prime.

Which Triangular Number is This?

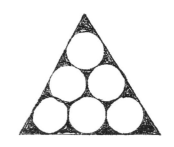

4a

It has three digits.
It is a triangle number.
It is palindromic.
The sum of its digits is prime.

4b

It is a three digit number.
It is a triangular number.
It is palindromic.
The sum of its digits is a perfect square.

How Many Triangles will Tessellate?

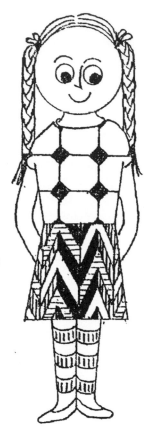

5a

How many triangles with the three sides measuring 3cm, 4cm and 5cm can you fit inside a rectangle measuring 0.3m by 0.6m?

5b

How many 1cm equilateral triangles can you fit inside a regular hexagon of side 5cm?

Tessa Lation
(She loves colouring patterns)

How Many Tiles are Needed?

6a How many square floor tiles of side 250mm are needed to tile the floor of a rectangular conservatory 4 metres by 3 metres?

6b How many rectangular floor tiles 300mm by 200mm are needed to tile a square conservatory of side 4.8 metres?

How Many Bricks are Needed?

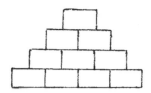

7a   A wall is built with layers of consecutive numbers of bricks, 1 on top, then 2, then 3 and so on. How many bricks are needed for a wall with twenty-eight layers?

7b   A wall is built with layers of consecutive odd numbers of bricks, 1 on top, then 3, then 5 and so on. How many bricks are needed for a wall with twenty-five layers?

How Many Cubes are Needed?

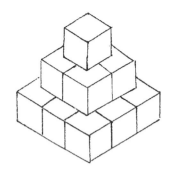

8a

A tower is built with layers of consecutive square numbers of cubes, 1 on top, then 4, then 9 and so on. How many cubes are needed for a tower with 20 layers?

8b

A tower is built with layers of consecutive square numbers of cubes, 1 on top, then 4, then 9 and so on. How many layers are there if the tower needs 9455 cubes?

How Many Sums of Money are There?

9a

I have in my purse one each of 1p, 2p, 5p, 10p and 20p coins. How many different sums of money is it possible to make from these coins?

9b

I have in my purse one each of 1p, 2p, 5p, 10p, 20p and 50p coins. How many different sums of money is it possible to make from these coins?

What is the Perimeter?

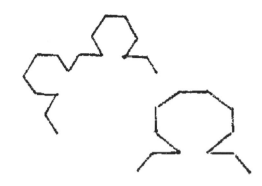

10a

What is the perimeter (in centimetres) of the shape formed when a regular hexagon of side 6cm has 6 regular hexagons each of side 2cm placed on the outside of it at the centre of each side as shown?

10b

What is the perimeter (in centimetres) of the shape formed when a regular octagon of side 5cm has 8 regular octagons each of side 2cm placed on the outside of it at the centre of each side as shown?

What is the Maximum Value?

11a Working through the alphabet in order, each letter is given a consecutive number from 1 to 26, but the 1 can start anywhere, Say: N = 1, O = 2, P = 3, ... M = 26. If the values of the letters of a word are then multiplied together what is the maximum value of the word CAT?

11b Working through the alphabet in order, each letter is given a consecutive number from 1 to 26, but the 1 can start anywhere, Say: F = 1, G = 2, H = 3, ... E = 26. If the values of the letters of a word are then multiplied together what is the maximum value of the word YAK?

How Many Ways can You Choose?

12a How many ways can Susan choose four of her ten friends to join her for her birthday treat?

12b A country dance needs lines of three people. How many ways can Ann choose the two dancers for her line from the other eleven people in the room?

How Much would You Pay?

13a Ellen, Fiona and Gemma each had 70p. Ellen bought a Pluto bar and a Shout bar and had 17p left. Fiona bought a Pluto bar and a Nutty bar and had 21p left. Gemma bought a Shout bar and a Nutty bar and had 18p left. How much would 3 Pluto bars, 4 Shout bars and 2 Nutty bars cost?

13b Henry, Ian and John each had 60p. Henry bought a banana and an apple and had 26p left. Ian bought an apple and a pear and had 29p left. John bought a banana and a pear and had 27p left. How much would 6 bananas, 3 apples and 5 pears cost?

How Long did He go Out For?

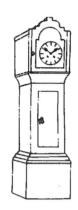

14a
Glancing at the reflection of Grandfather's clock in the mirror, John went out when he thought it was half past ten and returned when he thought it was quarter to nine on the same morning.
How many minutes was he really out for?

14b
Glancing at the reflection of Grandfather's clock in the mirror, Sam went out when he thought it was quarter to seven and returned when he thought it was half past two.
How many minutes was he really out for?

Which Multiple of 37 is This?

15a
It has three digits.
It is a multiple of 37.
It increases by 50% when
turned upside-down.

15b
It has three digits.
It is an even multiple of 37.
It stays the same when turned upside-down.

How Many Three Digit Numbers are There?

16a Steven's lucky number is 7. How many three digit numbers are there that contain at least one 7?

16b Oliver likes odd numbers. How many three digit numbers are there that contain only odd digits?

How Many Palindromic Dates are There?

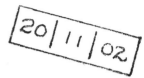

17a If the date is written with six digits, e.g.
1st Jan, 2002 as 01:01:02.
How many palindromic dates are there
between 1st Jan, 2002 and 31st Dec, 2999?

17b How many three digit palindromic numbers
are there?

How Many Leaves are There?

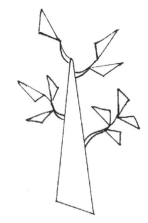

18a

The newly discovered triangular plant grows 1 leaf on day 1, 2 more leaves on day 2, 3 more on day 3, and so on. Assuming no leaves fall off, how many leaves will there be after 17 days?

18b

The newly discovered square plant grows 1 leaf on day 1, 4 more on day 2, 9 more on day 3, and so on. Assuming no leaves fall off, how many leaves will there be after 11 days?

How Many Cards are Sent?

Merry Xmas

19a

In a class of 20 pupils, 12 have mobile phones. Those without phones send cards to everyone in the class while the people with mobile phones send text messages to the others with phones and cards to the rest.

In all how many cards are sent?

19b

In a class of 25 pupils, 18 have mobile phones. Those without phones send cards to everyone in the class while the people with mobile phones send text messages to the others with phones and cards to the rest.

In all how many cards are sent?

How Many Ticks?

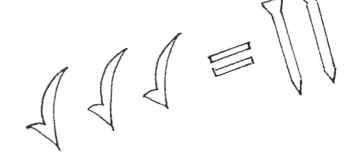

20a

3 ticks = 2 tacks
5 tacks = 4 tocks
7 tocks = 6 tucks
How many ticks make 240 tucks?

20b

2 ticks = 3 clicks
4 clicks = 5 flicks
3 flicks = 4 picks
How many ticks make 600 picks?

How Many Years?

21a   The year 2002 is one where the digits
add up to a total of 4.
How many times did this happen
between 1000 and 2000 A.D.?

21b   The year 2002 is palindromic with only
even digits.
How many times will this happen
between the years 2000 and 9999 A.D.?

What is the Smallest?

22a The sum of five consecutive even numbers is 230.
What is the smallest?

22b The sum of five consecutive odd numbers is 435.
What is the smallest?

How Many Space Diagonals?

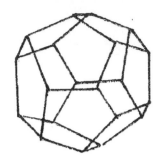

23a   A space diagonal joins two vertices that are not on the same face. How many space diagonals has a dodecahedron? (It has twelve pentagonal faces.)

23b   A space diagonal joins two vertices that are not on the same face. How many space diagonals has an icosahedron? (It has twenty triangular faces.)

What is the Mystery Number?

24a Find a four digit number such that the difference between its cube root and its square root is 48.

24b Find a three digit number such that the difference between its cube root squared and its square root cubed is 19 602.

How Long did They Live?

25a

Albert lived in the 18th century. His birthday was on 1st January in a year which was a perfect cube and he died on his birthday in a year which was a perfect square.

How long did he live?

25b

Annie was born in the 19th century. Her birthday was on 1st June in a year which was a perfect square and she died on 1st June in the 20th century in a year which was also a perfect square.

How long did she live?

How Old is Carol?

26a

Next year Barbara's age will be the difference between Carol's age next year and Avis's age two years ago.
Their present ages add up to 70 and Carol is the oldest.
How old is Carol?

26b

Doug, Eric and Fred all wrote down their present ages and they added up to 45.
They all rotated their present ages through 180° and then they added up to 183.
Fred is two years older than Eric and Doug's age is the same upside-down.
How old is Eric?

Which Multiple of 13 is This?

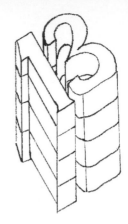

27a

It is a 3 digit number.
It is a multiple of 13.
It is palindromic.
The sum of its digits is the
number of fluid ounces in a pint.

27b

It is a 3 digit number.
It is a multiple of 13.
It is palindromic.
The sum of its digits is also
palindromic.

How Many
Reflecting Numbers?

28a

On a calculator display
5 is the reflection of 2.
How many three digit numbers display
reflective symmetry about a vertical line
through the centre digit?
The numbers cannot start with a zero.

28b

On a calculator display
5 is the reflection of 2.
How many four digit numbers display
reflective symmetry about a vertical line
between the two middle digits?
The numbers cannot start with a zero.

How Many Circles?

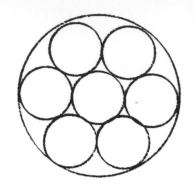

29a How many circles of radius 1cm
can be drawn inside a circle
of radius 5cm?

29b How many circles of radius 1cm
can be drawn inside a circle
of radius 7cm?

How Many Seats?

30a

An outside theatre is in the shape of a regular hexagon with the stage in the centre. The seats are banked up on all six sides with 50 seats on the top row of each side. There are 10 levels, each level having 3 less seats in each section than the one above. How many seats are there altogether?

30b

An outside theatre is in the shape of a regular octagon with the stage in the centre. The seats are banked up on all eight sides with 40 seats on the top row of each side. There are 8 levels, each level having 2 less seats in each section than the one above. How many seats are there altogether?

How Many Squares?

31a In a pack of Brownies, each 'Six' knitted an equal number of small squares, under 400. With these they made blankets. The Elves made a large square, the Imps four smaller squares, the Pixies a rectangle 27 squares long and the Gnomes a rectangle 36 squares long. How many squares did each 'Six' make?

31b Three students were given equal numbers of 1cm square tiles to make a mosaic. Harry made a square, Larry made a rectangle 54cm long and Gary a rectangle 72cm long. Harry worked out that he could have made twelve different rectangles instead of his square. How many tiles were they given?

How Many Tins?

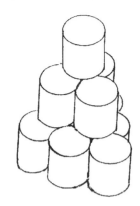

32a Julie works in a supermarket and likes displaying cans in pyramids. She makes a triangular based pyramid with one tin on top, then three, then six, then ten and so on. How many tins are needed for a pile ten layers high?

32b Jamie works in a supermarket and likes displaying cans in pyramids. He makes a square based pyramid with one tin on top, then four, then nine, then sixteen and so on. How many tins are needed for a pile twelve layers high?

Which Multiple of 11 is This?

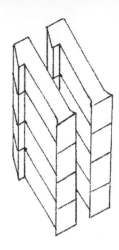

33a
It has three digits
It is a multiple of 11.
All the digits are odd.
The sum of its digits is
a perfect square.

33b
It has three digits
It is a multiple of 11.
All the digits are even.
It is a perfect square.

About How Many Sweets are There?

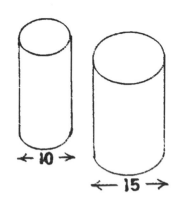

34a

Some small sweets, rather like smarties, are packed into cylindrical tubs of the same height. The tub of diameter 10cm holds about 800 sweets.
How many sweets would you expect in a tub of diameter 15cm?

34b

Some tiny sugar mice are packed into cylindrical tins of the same height. If the tin of diameter 12cm holds 192 mice about how many would you expect to find in a tin of diameter 9cm?

What is the Palindromic Total?

35a Bob lives in a street with 33 houses.
(33 is a palindromic number.)
He notes that the total sum of all the
numbers of all the even numbered houses
in the street is also palindromic.
What is it?

35b Anna lives in a street with 44 houses.
(44 is a palindromic number.)
She notes that the total sum of all the
numbers of all the odd numbered houses
in the street is also palindromic.
What is it?

What are the Values?

36a

Giving each letter of the alphabet a number from 26 to 1. (Z = 1, Y = 2, X = 3, ... A = 26) The values of the letters of a word are then multiplied together.
What are the values of PUN and WIT?

36b

Giving each letter of the alphabet a number from 26 to 1. (Z = 1, Y = 2, X = 3, ... A = 26) The values of the letters of a word are then multiplied together.
What are the values of WET and DRY?

What is the Eleventh Term?

37a The terms of a sequence are such that each is obtained from the previous one by adding 2 and then multiplying by −2. If the sequence begins with 2, what is the 11th term?

37b The terms of a sequence are such that each is obtained from the previous one by adding 3 and then multiplying by −2. If the sequence begins with 3, what is the 11th term?

How Many Days was He Away?

38a
Tom took off from Gatwick at 12 noon on 7th September, 1999 heading for America. He returned from his gap year at 12 noon on 8th August, 2000.
How many days was he away?

38b
Tim took off from Heathrow at 12 noon on 10th December, 1998 heading for Australia. He returned from his travels at 12 noon on 11th January, 2001.
How many days was he away?

Which Multiple of 19 is This?

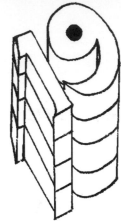

39a
It has three digits.
It is a multiple of 19.
It is palindromic.
All the digits are perfect squares.

39b
It has three digits.
It is a multiple of 19.
It is palindromic.
All the digits are even.

What do the Dates Add Up To?

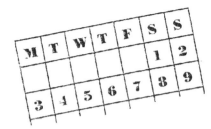

40a

On a calendar, each month is displayed in a block of seven days by four or five rows for the weeks. On his calendar John draws a square round a block of nine dates.

What is the maximum possible total for the numbers within this square?

40b

On a calendar, each month is displayed in a block of seven days by four or five rows for the weeks. On her calendar Mary draws a square round a block of nine dates.

What is the minimum possible total for the numbers within this square?

How Many Animals?

41a On the first day after the ark reached dry land, two animals left the ark. On the next day, four left, on day three, six left and so on. How many animals had left the ark after 36 days?

41b On day one there were 2 rabbits living on Paradise Island. The numbers doubled during each odd-numbered week and one died during each even-numbered week. After 15 weeks how many rabbits were living on the island?

What's My Rabbit Called?

42a

Starting with 1 give each letter of the alphabet
a consecutive number with alternate signs.
A = -1, B = 2, C = -3, D = 4,... Z = 26
Then multiply together the value of each of
the four letters in my rabbit's name.
The value is 429 so what's my rabbit called?

42b

Starting with 1 give each letter of the alphabet
a consecutive number with alternate signs.
A = -1, B = 2, C = -3, D = 4,... Z = 26
Then multiply together the value of each of
the five letters in my rabbit's name.
The value is -5915 so what's my rabbit
called?

What is the Least Number of Cubes?

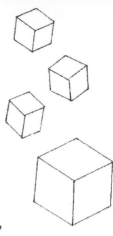

43a

Nayha has two sizes of cubes,
with sides of 3cm and 6cm.
What is the least number of cubes
needed to completely fill a box
measuring 24cm by 15cm by 15cm?

43b

Susan has two sizes of cubes,
with sides of 2cm and 4cm.
What is the least number of cubes needed
to completely fill a box measuring 24cm by
26cm by 22cm?

What is the New Area?

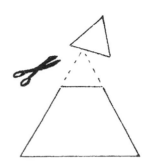

44a

An equilateral triangle of area 99cm^2 has three equilateral triangles cut from it, one at each corner. The resulting shape is a regular hexagon. What is the new area?

44b

An regular hexagon of area 72cm^2 has three equilateral triangles with sides the same length as the hexagon added to alternate sides. The resulting shape is an equilateral triangle. What is the new area?

Which Multiple of 59 is This?

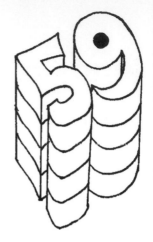

45a

It has four digits.
It is a multiple of 59.
All its digits are odd and
in ascending order.

45b

It has four digits.
It is a multiple of 59.
It has only two prime factors.
Its digits add up to 10.

How Many Chocolates?

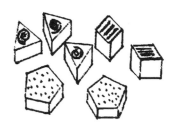

46a
Polly Gone worked in a chocolate factory
making sweets in the shapes of polygons.
She made a special box of chocolates with
3 triangular sweets, 4 square sweets,
5 pentagonal sweets and so on up to
10 decagonal sweets.
How many chocolates were there in the box?

46b
Holly decided to make a special Advent Gift
with chocolates in 24 numbered boxes.
She put two chocolates in the even numbered
boxes, three in those that were multiples of 3,
thus making 5 in those that were multiples of
6, except for number 24 into which she put
six. In all the rest she put a single chocolate.
How many chocolates were there altogether?

How Much was the Fish?

47a

Nafeesa bought fish, chips and mushy peas, her change from £5 was 70p. The difference between the price of the fish and the price of the chips was equal to the price of the mushy peas. How much did the fish cost?

47b

Carlos bought sweet and sour prawns, fried rice and a pancake roll, his change from £10 was £2.60. The difference between the price of the prawns and the price of the fried rice was equal to the price of the pancake roll.
How much did the prawns cost?

How Long was the Phone Call?

48a

In New York it is 7am when it is 12 noon in England. Susie in England phoned Billy Jo in New York. The call began at 6.45pm on Susie's clock and ended at 2.23pm on Billy Jo's clock.
How many minutes did the call last?

48b

In Tokyo it is 9pm when it is 12 noon in England. Henry in England phoned Satomi in Tokyo. The call began at 10.30am on Henry's clock and ended at 8.17pm on Satomi's clock.
How many minutes did the call last?

How Many Separate Digit Cards?

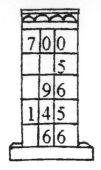

49a A Community Centre has a weekly raffle with exactly 999 tickets (1 to 999). The 5 winning numbers are displayed, each of the digits on a separate card. If numbers below 100 do not need zeros in front and sixes can be used as nines what is the least number of cards needed to show all possible winning numbers?

49b A Social Club has a weekly raffle with exactly 999 tickets (1 to 999). The 6 winning numbers are displayed, each of the digits on a separate card. If numbers below 100 do not need zeros in front and sixes can be used as nines what is the least number of cards needed to show all possible winning numbers?

What is the Value of Table?

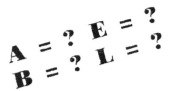

A = ? E = ?
B = ? L = ?

50a

Each letter in these words is given a number below 10 and the values of the letters in each word are then multiplied together. Given that
BAT = 90, LET = 168 and BET = 105
what is the value of TABLE?

50b

Each letter in these words is given a number below 10 and the values of the letters in each word are then multiplied together. Given that
RAT = 40, RAG = 60, GET = 216 and
GATE = 432 what is the value of GREAT?

How Many go by Bus?

51a In a school of 1200 pupils, 580 are girls.
One third of the pupils travel to school by
bus. 462 boys do not go by bus.
How many girls do go by bus?

51b In a factory of 2000 workers, 1150 are
female. Two fifths of the workers travel to
work by bus. 595 males do not go by bus.
How many females do go by bus?

How Many were there Altogether?

52a

Sam and Ben collect marbles.
Sam says to Ben 'If you give me 5
marbles, I'll have twice as many as you'.
Ben says 'If you give me 4 we would have
the same number'.
How many marbles were there altogether?

52b

Sally and Susie collect stamps.
Sally says to Susie 'If you give me 10 stamps,
I'll have three times as many as you'.
Susie says 'If you give me 100 we would
have the same number'.
How many stamps were there altogether?

What's the Total?

$$101 + 102 + \ldots \ldots + 105 = ?$$

53a What is the total of 5 consecutive three-digit whole numbers where the first is a perfect square and the last is a perfect cube?

53b What is the total of 10 consecutive three-digit whole numbers where the first is a perfect cube and the last is a perfect square?

What is the Minimum Number?

54a At a party half the guests are men. The number of men with bow ties is 3 less than the number of women wearing black dresses. There are 3 couples where the man has a bow tie and the woman is wearing a black dress. At least 6 ladies are not in black.

54b In my train compartment, half the passengers are smoking. The number reading newspapers is 3 more than the number using mobile phones. I am in a carriage using my mobile. Two of the smokers are reading a paper and no-one else is doing more than one thing. What is the minimum number of passengers?

Which Multiple of 18 is This?

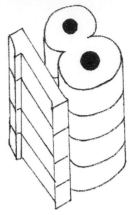

55a   It has three digits.
It is a multiple of 18.
It has three different digits in ascending order.
Whichever way its three digits are arranged it
is still a multiple of 18.

55b   It has three digits.
It is a multiple of 18.
It is palindromic.
The product of its digits is 128.

What Number was the Middle House?

56a Five houses next door to each other were on the odd side of the road. They were all two digit numbers and in ascending order, the fourth one was the only prime.
What number was the middle house?

56b Five houses next door to each other were on the even side of the road. They were all three digit numbers and the largest was a perfect cube. The second lowest was a triangle number.
What number was the middle house?

How Many People were at the Picnic?

57a

I took a photo of some friends on a picnic. Six of them had crisps, eight of them a drink and five were eating sandwiches. Some had one item and some two. What is the least number of people at this picnic?

57b

I went on a picnic with my classmates. Half were boys and a quarter of everyone there was older than me. I was the third oldest girl and there were nine boys younger than me. What is the least number of people at this picnic?

What is the Surface Area?

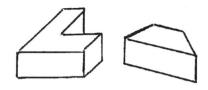

58a

A polystyrene moulding has a cross section in the shape of a letter L with its longer edges 10cm and all other measurements 5cm, including its depth.
What is its surface area?

58b

A polystyrene moulding has a cross section in the shape of an isosceles trapezium with sides 15cm, 5cm, 9cm and 5cm. Its depth is 10cm.
What is its surface area?

How Many Different Ways?

59a Five people, Alice, Ben, Carol, David and Ellen sit in a line on five chairs.
How many different ways can they arrange themselves if the two boys are not allowed to sit together?

59b How many different arrangements are there of the letters in the word HARPY?
How many different arrangements are there of the letters in the word HAPPY?

How Many can be Fitted In?

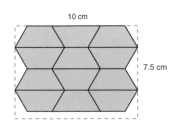

10 cm

7.5 cm

60a

The design above includes isosceles trapeziums with sides 2cm, 2cm, 2cm and 4cm. Within a rectangle 10cm by 7.5cm, just twelve whole trapeziums can be fitted. How many whole trapeziums can be fitted into a square of side 30cm?

60b

This design includes regular hexagons of side 2cm. Within a rectangle 10cm by 7.5cm, five whole hexagons can be fitted.

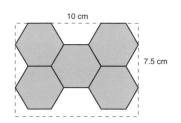

10 cm

7.5 cm

How many whole hexagons can be fitted into a square of side 30cm?

What is the Shortest Distance?

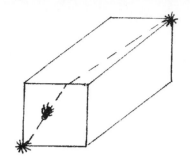

61a What is the shortest distance an ant could crawl, on the surface of a box, to reach from one bottom corner to the opposite top corner? The box measures 15cm by 10cm by 10cm. Hint: draw a net of the box

61b What is the shortest distance an ant could crawl, on the surface of a box, to reach from one bottom corner to the opposite top corner? The box measures 24cm by 16cm by 16cm. Hint: draw a net of the box

How Many Slabs?

62a

Darren decided to slab the patio with a diagonal pattern of square slabs. In a rectangle 1.8m by 1.2m eight whole slabs are needed. The left over triangles would be filled with pebbles or plants. How many slabs are needed for a patio 2.4m by 3.6m?

62b

Wayne decided to slab the patio with a diagonal pattern of square slabs. In a rectangle 1.8m by 1.2m eight whole slabs are needed. The left over triangles would be filled with pebbles or plants. How many slabs are needed for a patio 3.6m by 3.6m?

How Long will it Take?

63a Sir Cumference has a new car which he drives around the estate where he lives. The estate has a perfectly circular ring road of length 3770m and several straight roads across the middle. How long will it take him (calculated to the nearest second) to drive across the middle at 36 km/h?

63b Sir Cumference has a new car which he drives around the estate where he lives. The estate has a perfectly circular ring road of length 3927m and several straight roads across the middle. How long will it take him (calculated to the nearest second) to drive across the middle at 25 km/h?

What is the Mystery Number?

64a It has four digits.
It is a perfect square of a prime number.
It reads the same upside-down.

64b It has four digits.
It is a perfect square.
If you reverse the first two digits you get
another smaller perfect square.

What is the 100th Term?

OO 2

&& 5

&&& 9

&&&& 14

65a
What is the 100th term
of the sequence

2, 5, 9, 14, 20, ... ?

65b
What is the 100th term
of the sequence

1, 5, 11, 19, 29, ... ?

What is their Number?

66a The Adams, the Browns and the Clarks live in three houses next door to each other with the Browns in the middle and the Adams at the house with the smallest number. The house numbers differ by 2 and all three add up to 339. What is the Clarks' number?

66b In the Paradise Hotel, room numbers are identical on each floor except for the first number, e,g. Room 214 is directly above 114 and below 314. The Darwins, the Newtons and the Pascals stay in rooms above each other with the Newtons in the middle. The room numbers add up to 963.
What is the Newtons' number?

How Many Sweets?

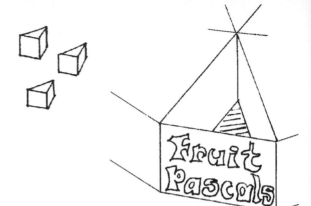

67a Fruit Pascals are triangular sweets in the shape of a prism, the cross section being an equilateral triangle of side 1cm and the depth is also 1cm. How many sweets could be packed into a hexagonal box of side 3cm and depth 5cm which has six triangular dividers?

67b Fruit Pascals are triangular sweets in the shape of a prism, the cross section being an equilateral triangle of side 1cm and the depth is also 1cm. How many sweets could be packed into a hexagonal box of side 4cm and depth 2cm which has six triangular dividers?

How Old were the Grandparents?

68a

In 1936 Grandma was twice
as old as Sally was in 1999.
In 1944 Grandma was three times
as old as Sally was in 1999.
How old was Grandma in 1999?

68b

In 1921 Grandpa was twice
as old as Tom was in 2000.
In 1930 Grandpa was three times
as old as Tom was in 2000.
How old was Grandpa in 2000?

How Many Squares are There?

69a

How many squares are there
on a chess board?
Hint: There are 64 of the smallest size.

69b

How many squares are there
on a scrabble board?
Hint: There are 225 of the smallest size.

What is the Smallest Number?

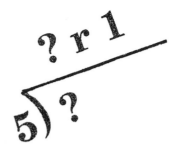

70a Find the smallest number such that
when it is divided by 5 or 7 or 8,
the remainder each time is 1.

70b Find the smallest number such that
when it is divided by 6 or 7 or 11,
the remainder each time is 1.

What's the Total?

$$12 + 13 + \ldots$$
$$\ldots + 20 + 21 = ?$$

71a What is the total of ten consecutive two digit whole numbers where the first and last are perfect squares?

71b What is the total of ten consecutive two digit whole numbers where the first is a perfect cube and the last a perfect square?

How Much Altogether?

72a Ann, Betty and Carol are counting their money.
Ann and Betty together have £5.98.
Betty and Carol together have £6.47.
Ann and Carol together have £8.51.
How much money do they have altogether?

72b Adam, Ben and Carl are weighing themselves.
Together Adam and Ben weigh 102kg.
Ben and Carl weigh 97kg.
Adam and Carl weigh 93kg.
What is their total weight in kg?

How Many Lines are There?

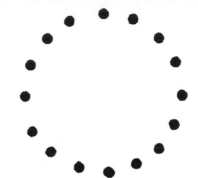

73a Twenty four dots are spaced evenly in a circle. If every dot is joined to every other dot with a straight line, how many lines are there?

73b Twenty eight dots are spaced evenly in a circle. If every dot is joined to every other dot with a straight line, how many lines are there?

How Many Tabs are There?

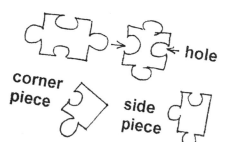

corner piece

side piece

hole

74a

A rectangular Jigsaw puzzle is divided up into pieces each 2cm by 2cm. They are joined to each other by almost circular tabs at the centre of each edge which fit into similar circular holes. Outside edges are straight.

In a jigsaw 10cm by 8cm how many tabs are there?

74b

A rectangular Jigsaw puzzle is divided up into pieces each 2cm by 2cm. They are joined to each other by almost circular tabs at the centre of each edge which fit into similar circular holes. Outside edges are straight.

In a jigsaw 16cm by 12cm how many tabs are there?

What Number did I Start With?

$$\sqrt{\frac{?}{2}} \times 4 \times 9 = ?$$

75a

I started with a three-digit number and divided it by 2. I then took the square root of the result. I multiplied this by 4 and then by 9 and found that I had got back to the number that I started with.

What number did I start with?

75b

I started with a three-digit number and divided it by 2. I then took the cube root of the result. I multiplied this by 7 and then by 14 and found that I had got back to the number that I started with.

What number did I start with?

What is the Maximum Value?

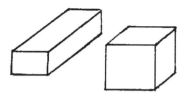

76a What is the maximum volume in cubic centimetres of a cuboid whose length, width and height are all whole numbers of centimetres, given that these three numbers add up to 20?

76b What is the maximum surface area in square centimetres of a cuboid whose length, width and height are all whole numbers of centimetres, given that these three numbers add up to 22?

How Many will Fit?

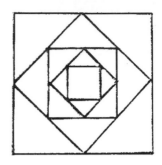

77a
Inside a square, another square is drawn by joining the mid points of the sides. This is then repeated until there are five squares.

How many times does the smallest square fit into the original?

77b
Inside an equilateral triangle, another triangle is drawn by joining the mid points of the sides. This is then repeated until there are five triangles drawn.

How many times does the smallest triangle fit into the original?

How Many Sides does this Polygon Have?

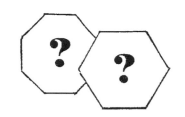

78a

This regular polygon has an odd number of sides, less than 100. The sum of its interior angles is a four digit perfect square.

How many sides does this polygon have?

78b

This regular polygon has an even number of sides, less than 100. The sum of its interior angles is a four digit perfect square.

How many sides does this polygon have?

What was the Door Code?

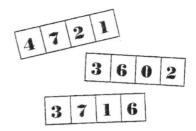

79a

To help pupils remember the school's four digit security code the history teacher said it was the year of the 'Battle of Bosworth Field'. The maths teacher said it was easy to remember because it was the 54th triangle number. What was the door code?

79b

To help pupils remember the school's four digit security code the history teacher said it was the year of the death of Sir Francis Drake. The maths teacher said it was easy to remember because it was the 56th triangle number. What was the door code?

How Many Right Angles are There?

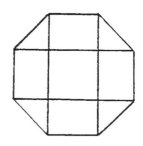

80a

Start with a regular octagon. Then draw all the diagonals that are perpendicular to the sides. How many right angles are there in the completed diagram?

80b

Start with a regular hexagon and draw in all the diagonals. How many right angles are there in the completed diagram?

How Many Zog Units are There?

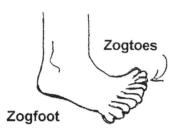

Zogtoes

Zogfoot

81a

On planet Zog there are 100 zogminutes in a zoghour, 20 zoghours in a zogday and 10 zogdays in a zogweek.
The days of the week are called Oneday, Twoday, Threeday etc.
How many zogminutes are there from mid-day on Twoday until midnight on Fourday?

81b

On planet Zog there are10 zogtoes in a zogfoot, 5 zogfeet in a zogyard and 2000 zogyards in a zogmile.
How many zogtoes are there in three quarters of a zogmile?

How Much Profit was Made?

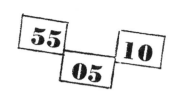

82a

One hundred tombola tickets were sold, numbered 1 to 100, at 25p each. Each ticket which included a '5' won a prize costing 20p and each ticket ending in '0' won a prize costing 50p. The prize for ticket 50 cost £2.00

How much profit was made?

82b

In the 'Guess the number of Sweets in a Jar' competition, 148 people had a go at 20p each and a quarter of them also gave a donation of 10p. There was one winner who won the jar of 132 sweets which cost an average of 7p each.

How much profit was made?

How Many
Half-time Scores?

7-5

11-7

83a

The final score of the hockey match was 7-5.

How many possible different half-time scores could there have been?

83b

The final score of the netball match was 11-7.

How many possible different half-time scores could there have been?

How Long was the Flight?

84a   Simon took off from Gatwick at 10.16am local time and landed at Houston at 1.15pm local time. If Houston is 6 hours behind England, how long did the flight last?

84b   Karen took off from Birmingham at 9.08am local time and landed at Chicago at 11.55am local time. If Chicago is 6 hours behind England, how long did the flight last?

What is the Maximum Number of Spots?

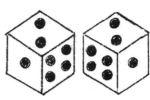

85a Six-sided dice have their spots arranged so that opposite faces add up to 7. If ten dice are arranged in a vertical tower on a solid table, what is the maximum number of spots that could be visible from one position?

85b Six-sided dice have their spots arranged so that opposite faces add up to 7. If eight dice are arranged in a 2 x 2 x 2 cube on a solid table, what is the maximum number of spots that could be visible from one position?

What is the Perimeter?

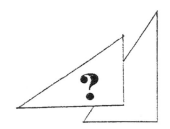

86a

Ignoring the different units, the perimeter of this right angled triangle is numerically half its area. In addition the length of its shortest side is an odd perfect square and the other two sides are consecutive two-digit whole numbers.

What is its perimeter?

86b

Ignoring the different units, the perimeter of this right angled triangle is numerically a third of its area. In addition the length of its shortest side is a prime number and the other two sides are consecutive two-digit whole numbers.

What is its perimeter?

How Many Ornaments are There?

87a

Victoria collects china dogs. One third of her collection stand up on all 4 legs, the rest either beg on 2 legs or lie down. The number that lie down exceeds the number that beg by half the number that stand.

If 21 dogs beg, how many dogs are there?

87b

Corinne collects china cats which either stand on 4 legs, on 2 legs or lie down. Half lie down and one third stand on 4 legs.

If she had 6 more that stood on two legs, that number would be a quarter of the number that she now has.

How many cats are there?

What is the Minimum Number of Bricks?

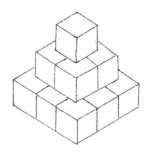

88a A square-based pyramid is built from cubical bricks, one on the top layer, four on the next, nine on the next, 16 on the next and so on.

What is the minimum number of bricks needed if the pyramid is to be dismantled and rebuilt into two separate cubes?

88b A square-based pyramid is built from cubical bricks, one on the top layer, four on the next, nine on the next, 16 on the next and so on.

What is the minimum number of bricks needed if the pyramid is to be dismantled and rebuilt into three separate cubes?

Which Binary Number is This?

101010I

11011

11010101I

101101101

89a

Which is the lowest multiple of 15 which when written as an eight digit binary number reads the same forwards as backwards?

89b

Which is the lowest multiple of 13 which when written as an eight digit binary number reads the same forwards as backwards?

How Many Days?

90a

Trixie Triangle was learning to play the piano. She decided to practise for 5 minutes on day 1, 15 minutes on day 2, 25 minutes on day 3, and so on.

How many days would it be before she was practising for more than half the day?

90b

Sam Square went on a gap year, round the world trip. He stayed 1 day at the first place, 4 days at the second, 9 days at the third and so on. He visited ten places in all.

For how many days of the second year was he away from home?

Answers

1a 216	1b 1728	26a 34	26b 16
2a 689	2b 901	27a 767	27b 949
3a 169	3b 676	28a 12	28b 20
4a 595	4b 171	29a 19	29b 37
5a 300	5b 150	30a 2190	30b 2112
6a 192	6b 384	31a 324	31b 1296
7a 406	7b 625	32a 220	32b 650
8a 2870	8b 30	33a 979	33b 484
9a 31	9b 63	34a ~1800	34b ~108
10a 84cm	10b 136cm	35a 272	35b 484
11a 10608	11b 6240	36a 858, 504	36b 616, 414
12a 210	12b 55	37a 3412	37b 5118
13a £2.35	13b £2.31	38a 336	38b 763
14a 105	14b 255	39a 494	39b 646
15a 666	15b 888	40a 207	40b 81
16a 252	16b 125	41a 1332	41b 258
17a 29	17b 90	42a MACK	42b GEMMA
18a 153	18b 506	43a 88	43b 456
19a 248	19b 294	44a 66cm^2	44b 108cm^2
20a 525	20b 240	45a 1357	45b 1711
21a 10	21b 20	46a 52	44b 57
22a 42	22b 83	47a £2.15	45b £3.70
23a 100	23b 36	48a 38	48b 47
24a 4096	24b 729	49a 102	49b 121
25a 36	25b 87	50a 5040	50b 2160

Answers

51a	242	51b	545
52a	54	52b	440
53a	615	53b	2205
54a	24	54b	6
55a	468	55b	828
56a	95	56b	212
57a	11	57b	28
58a	350cm^2	58b	436cm^2
59a	72	59b	120, 60
60a	153	60b	72
61a	25cm	61b	40cm
62a	39	62b	61
63a	120	63b	180
64a	6889	64b	9216
65a	5150	65b	10 099
66a	115	66b	321
67a	270	67b	192
68a	79	68b	97
69a	204	69b	1240
70a	281	70b	464
71a	205	71b	315
72a	£10.48	72b	146kg
73a	276	73b	378
74a	31	74b	82
75a	648	75b	686

76a	294	76b	322
77a	16	77b	256
78a	47	78b	22
79a	1485	79b	1596
80a	48	80b	36
81a	5000	81b	75 000
82a	£14.90	82b	£24.06
83a	48	83b	96
84a	8h.59m	84b	8h.47m
85a	114	85b	66
86a	90	86b	182
87a	84	87b	72
88a	91	88b	55
89a	165	89b	195
	10100101		11 000 011
90a	72	90b	20 or 19
			(leap year)

Using Sketches and Clip Art

In these more visual days, it is commonplace to see simple sketches being used to enliven a page and children do rather expect it. Often these illustrations have only a passing connection or reference to the topic or puzzle being offered. A quick sketch is all that is needed or if not, there are masses of clip art illustrations available in books or as computer files. You can then make a selection for your daily puzzles.

If you have enjoyed this book there may be other Tarquin books which would interest you, including 'Mathematical Treasure-Hunts' by Vivien Lucas, 'Maths Snacks' by Jon Millington and several other mathematical puzzle books. Tarquin books are available from bookshops, toy shops and gift shops or in case of difficulty, directly by post from the publishers.
For an up-to-date catalogue please write to info@tarquingroup.com, phone (0044) (0) 1727 833866 or write to us at Tarquin Publications Suite 74, 17 Holywell Hill, St Albans, AL1 1DT Alternatively, see us on the Internet at http://www.tarquin-books.demon.co.uk